Omicron Sub Variants – A Master Of Disguise: A guide On What We Need To Know

Andrea Raymond

TABLE OF CONTENTS

INTRODUCTION

There are various sub-variations of Omicron, all designated as Omicron, and variants of concern that the World Health Organization is following. The ones focusing on right now are mainly BA.4, BA.5, and BA.2.75.

The omicron subvariant known as BA.5 was first found in South Africa in February 2022 and spread fast around the globe. As of the second week of July 2022, BA.5 formed roughly 80% of COVID-19 versions in the United States.

Soon after researchers in South Africa reported the original version of the omicron variant (B.1.1.529) on Nov. 24, 2021, many scientists speculated that if omicron's numerous mutations made it either more transmissible or better at immune evasion than the preceding delta variant, omicron could become the dominant variant around the world.

The omicron variety did indeed become prevalent early in 2022, and numerous sublineages, or subvariants, of omicron, have subsequently emerged: BA.1, BA.2, BA.4, and BA.5, among others. With the continuous development of such highly transmissible variations, it is obvious that SARS-CoV-2, the virus that causes COVID-19, is efficiently exploiting basic tactics that viruses utilize to bypass the immune system. These escape techniques vary from altering the structure of critical proteins identified by your immune system's protective antibodies to camouflaging its genetic material to mislead human cells into believing it is a part of themselves instead of an enemy to fight.

BA.5 is one of the newest subvariants of the original Omicron lineage. There is a unique cluster of mutations on the spike protein, which has caused the BA.5 subvariant to possibly be the most infectious strain of

SARS-CoV-2 thus far. But it does not seem to contribute to more severe consequences than other COVID-19 variants and subvariants.

The omicron subvariant of COVID-19, BA.5, has become one of the dominant strains of the virus in the U.S. It's the most readily distributed strain to date and is able to avoid protection from COVID infection and vaccination.

Chapter 1: The Measurement Of Virus Transmissibility In A Population And The Evolution Of Omicron Sublineages.

The basic reproduction number, Ro – pronounced “R-naught” – measures the transmissibility of a virus in a yet-uninfected population.

Once a percentage of people in a community become immune owing to earlier infection or vaccination, epidemiologists use the term effective reproduction number, denoted Re or Rt, to quantify the transmissibility of the virus. The Re of the omicron form has been predicted to be 2.5 times higher than the delta variant. This improved transmissibility most likely enabled omicron out-compete delta to become the dominant variation.

Most people assume that being outdoors is safer than being inside, but there are still certain concerns to bear in mind while the BA.5 Omicron subvariant continues to propagate. Health specialists have recognized that outdoor transmission of SARS-CoV-2 has happened, but it's more probable that the virus may be caught inside in poorly ventilated places. While outdoor activities are fundamentally safer than inside ones, the heightened transmissibility of BA.5 will certainly lead to spread in both situations. One point to consider regarding outdoor transmission is that virus particles might disseminate more rapidly owing to wind speed and air currents. With that, certain outdoor environments, such as a closely packed outdoor performance, might potentially boost a person's risk of developing SARS-CoV-2.

The broader issue, therefore, is what is driving the development of omicron sublineages.

The solution to it is a well-known mechanism called natural selection. Natural selection is an evolutionary process where features that offer a species a reproductive advantage continue to be handed down to the next generation, while ones that don't are phased out via competition. As SARS-CoV-2 continues to spread, natural selection will favor mutations that provide the virus the largest survival benefit. BA.5 has a growth advantage over the other sub-lineages that have been found. For severity, there is no increase in severity of those infected with BA.5 compared to the other sub-lineages BA.1 and BA.2 for example.

As part of its ongoing effort to monitor variations, WHO's Technical Advisory Group on SARS-CoV-2 Virus Evolution (TAG-VE) met recently to examine the newest information on the Omicron variant of concern, particularly its sublineages BA.1 and BA.2. Based on existing evidence regarding transmission, severity, reinfection, diagnostics, treatments, and effects of vaccines, the panel underlined that the BA.2 sublineage should continue to be regarded a variant of concern and that it should be classed as Omicron. The organization highlighted that BA.2 should continue to be monitored as a unique sublineage of Omicron by public health authorities.

The Omicron variation of concern is presently the prevalent variant circulating worldwide, accounting for virtually all sequences submitted to GISAID. Omicron is made up of multiple sublineages, each of them being monitored by WHO and partners. Of these, the most prevalent ones are BA.1, BA.1.1 (or Nextstrain clade 21K) and BA.2 (or Nextstrain clade 21L) (or Nextstrain clade 21L). At a worldwide level, the fraction of reported sequences classified BA.2 has been growing compared to BA.1 in recent weeks, while the global circulation of all variants is allegedly dropping. BA.2 varies from BA.1 in its genetic sequence, including minor amino acid changes in the spike protein and other proteins.

Studies have indicated that BA.2 has a growth advantage over BA.1. Studies are continuing to determine the causes for this growth advantage, but early results show that BA.2 is fundamentally more transmissible than BA.1, which presently remains the most prevalent Omicron sublineage recorded. This variation in transmissibility seems to be significantly lower than, for example, the difference between BA.1 and Delta. Further, while BA.2 sequences are growing in percentage compared to other Omicron sublineages (BA.1 and BA.1.1), but there is still a documented drop in total cases worldwide. Studies are examining the probability of reinfection with BA.2 compared to BA.1. Reinfection with BA.2 after infection with BA.1 has been observed, however, early results from population-level reinfection studies show that infection with BA.1 gives good protection against reinfection with BA.2, at least for the brief duration for which data are available.

While reaching the aforementioned judgment, the TAG-VE also reviewed early laboratory evidence from Japan created using animal models without any immunity to SARS-CoV-2 which revealed that BA.2 may cause more severe sickness in hamsters compared to BA.1. They also evaluated real-world data on clinical severity from South Africa, the United Kingdom, and Denmark, where protection from vaccination or spontaneous infection is high: in this data, there was no reported difference in severity between BA.2 and BA.1.

WHO has decided to continue to carefully monitor the BA.2 lineage as part of Omicron and encourages nations to continue to remain alert, to monitor and report sequences, as well as to perform independent and comparative investigations of the numerous Omicron sublineages. The TAG-VE meets periodically and continues to review available data on the transmissibility and severity of variations, and their influence on diagnostics and treatments.

Chapter 2: The Spread Of Omicron And Its Offshoots

All viruses gain new mutations over time, and SARS-CoV-2 is no exception. Some of these modifications do not influence the virus's behavior, whilst others may modify features such as how infectious it is, the tissues it infects, or the severity of the sickness it produces. A collection of closely related variations that share a common ancestor is termed a lineage, and these may then branch off into sub-lineages, which is what presently seems to be occurring with Omicron.

The same problem occurred with prior variations of concern. For instance, the Delta variation included more than 200 sub-lineages until Omicron started to take control. So far, the Omicron variation is assumed to have been divided into three sub-lineages – BA.1, BA.2, and BA.3 – and these will continue to change in the future.
According to the CDC COVID tracker, BA.5 is the prevalent strain of COVID-19. It is more readily transmitted than other strains because it evades protection from prior COVID-19 infection and vaccination. That implies that even if you were infected with delta or omicron BA.1, you may still develop BA.5. Your past immunity does not protect you against the newest strain. The alterations of the BA.5 subvariant have

enabled it to get past the body's immune system more readily. This implies whether you had prior SARS-CoV-2 infection or have been vaccinated and boosted against COVID-19, you are not entirely protected. Compared with the BA.2 subvariant, BA.5 has been able to spread more readily because of its capacity to bypass immunological protection against infection caused by past infection or vaccination. This is particularly if protection has diminished over time, resulting in more frequent breakthrough infections and reinfections. While breakthrough infections are not resulting in severe disease, each reinfection does carry the chance of developing protracted COVID, also known as post-COVID disorders.

Several processes contribute to the higher transmissibility of SARS-CoV-2 mutations. One is the capacity to bind more firmly to the ACE2 receptor, a protein in the body that typically helps control blood pressure but may also assist SARS-CoV-2 to penetrate cells. The more recent omicron sublineages feature alterations that make them better at avoiding antibody protection while keeping their capacity to successfully connect to ACE2 receptors. The BA.5 sublineage may avoid antibodies from both immunization and past infection. Omicron sublineages BA.4 and BA.5 share multiple mutations with previous omicron sub-lineages but also feature three distinct mutations: L452R, F486V, and reversal (or the absence of mutation) of R493Q. L452R and F486V in the spike protein let BA.5 avoid antibodies. In addition, the L452R mutation helps the virus attach more efficiently to the membrane of its host cell, a critical characteristic related to COVID-19 disease severity.

When a virus is spreading extensively and causing multiple illnesses, the risk of the virus changing rises. The more possibilities a virus has to disseminate, the more opportunities it has to undergo alterations. Omicron has now been found in most nations, after the variation was first detected in November 2021 and there is a lower risk of hospitalization for Omicron compared to the Delta form. But WHO advises that it should not be disregarded as “mild”. An increase in the frequency of COVID-19-associated fatalities due to the Omicron variation has been documented in numerous countries, particularly when immunization levels are low among susceptible groups.

It is vital to remember that all variations of COVID-19 may cause serious sickness or death, which is why controlling the transmission of the virus and lowering your risk of exposure to the virus is so important.

Chapter 3: Symptoms Of Omicron BA.5

Most persons who test positive for any variation of COVID-19 often have some symptoms for a few weeks. People who have extended COVID-19 symptoms may endure health difficulties for four or more weeks after initially being infected, according to the CDC.

BA.5 symptoms are identical to prior COVID-19 variants and sub-variants.

The most prevalent symptoms include:

- fever
- Runny nose
- Coughing
- Sore throat
- Headache
- Muscle pain
- Fatigue.

Additionally, the prevalence and severity of a patient's symptoms might be altered by their COVID-19 vaccination status, other health problems, age, and history of past infection, according to the Centers for Disease Control and Prevention (CDC).

Even persons who have partial protection from a prior illness or immunization might still have a breakthrough infection. Breakthrough infections occur in patients who have been immunized or previously had COVID. However, the majority of breakthrough infections are not producing serious sickness, as contrasted to early in the pandemic when no one had immunity.

A new study reveals that with each recurrent COVID infection – even silent illness – the chance for consequences rises. These include:

- Stroke
- Heart attack
- Diabetes
- Digestive and kidney disorders
- Long-term cognitive impairment, including Dementia.

Each reinfection also carries with it the risk of long COVID or ongoing COVID symptoms that can last for weeks or months after infection

Chapter 4: The Next Step

People need to realize that variations like Omicron and BA.5 are a normal part of the development of the virus and they aren't unexpected. Delta was never meant to be the final version and Omicron is not going to be the last one neither will BA.5 be the end game. As long as there is a COVID-19 epidemic somewhere in the globe, there is going to be something new that emerges."

As the virus continues to spread, this evolutionary tendency will likely lead to the formation of more transmissible varieties that are capable of immunological escape. While it is impossible to forecast what variations will come next, researchers cannot rule out the potential that some of these mutations might contribute to greater disease severity and higher hospitalization rates. As the virus continues to mutate, most individuals will get COVID-19 several times despite vaccination status. This might be perplexing and irritating for some and may add to vaccination reluctance. Therefore, it is crucial to know that vaccines protect you against serious sickness and death, not necessarily from being infected. Research over the last two and a half years has helped scientists understand a lot about this novel virus. However, many unsolved concerns remain since the virus continually develops, and is like attempting to aim a constantly shifting goalpost. While updating vaccines to match circulating strains is a possibility, it may not be practicable in the near term since the virus develops too fast. Vaccines that generate antibodies against a wide spectrum of SARS-CoV-2 variants and a combination of broad-ranging therapies, including monoclonal antibodies and antiviral medicines, will be important in the battle against COVID-19.

Chapter 5: Vaccines For Preventing Omicron BA.5

In the UK, three vaccinations are now in use, after regulatory authorization. These are the BNT162b2 vaccine, created by Pfizer and BioNTech, the ChAdOx1 nCoV-19 (AZD1222) vaccine, developed by the University of Oxford and AstraZeneca, and the mRNA-1273 vaccine developed by Moderna. These vaccines have been licensed for emergency use by the Medicines and Healthcare products Regulatory Agency (MHRA) (MHRA).

The BNT162b2 and the mRNA-1273 vaccines are COVID-19 mRNA vaccines produced by Pfizer and BioNTech and Moderna, respectively. They individually carry the genetic code (mRNA) of the spike protein, which is located on the surface of the SARS-CoV-2 virus. Once within the body, the spike protein is generated, prompting the immune system to detect it and begin an immunological reaction.

This implies that if the body subsequently meets the spike protein of the coronavirus, the immune system will detect it and eliminate it before causing illness. As there is no full or live virus involved, the vaccinations cannot induce COVID-19 illness. The mRNA is spontaneously destroyed after a few days.

The safety and effectiveness of the Pfizer-BioNTech vaccine have been examined in clinical studies of approximately 44,000 persons in six countries: the USA, Germany, Brazil, Argentina, South Africa, and Turkey. The study indicated that vaccination can prevent 95% of COVID-19 cases. This indicates that in a group of 20 persons who are vaccinated if all are exposed to the coronavirus, 19 people will be protected from having COVID-19. The experiment also revealed that the vaccination gives equivalent protection to persons of various ages, races, and ethnicities.

The safety and effectiveness of the Moderna vaccine were evaluated in over 30,000 individuals throughout the United States, including elderly adults, persons from ethnic minorities, and those with underlying health issues. The vaccination was demonstrated to prevent 94% of COVID-19 instances and prevented all severe cases of COVID-19.

Oxford-AstraZeneca ChAdOx1 nCoV-19 (AZD1222) (AZD1222)

The ChAdOx1 nCoV-19 vaccine was developed by the University of Oxford and AstraZeneca. The vaccine works by delivering the genetic code of the SARS-CoV-2 spike protein to the body's cells, comparable to the mRNA vaccines. Once within the body, the spike protein is generated, prompting the immune system to detect it and begin an immunological reaction. This implies that if the body subsequently meets the spike protein of the coronavirus, the immune system will detect it and eliminate it before causing illness.

This Oxford-AstraZeneca vaccine leverages the ChAdOx1 technology, which has been developed and improved by the Jenner Institute over the previous 10 years. This sort of vaccination technology has been tried for many different illnesses such as influenza (flu) and middle east respiratory syndrome (MERS), another form of coronavirus.
The ChAdOx1 nCoV-19 vaccine has been evaluated by the University of Oxford in clinical trials of over 23,000 patients in the UK, Brazil, and South Africa. AstraZeneca is also launching an additional study with 40,000 participants in the USA, Argentina, Chile, Colombia, and Peru.

Interim findings from the UK and Brazil studies revealed that vaccination can prevent 70.4% of COVID-19 infections. This was determined across two distinct groups of patients, who got two different dosing regimens. The vaccination was demonstrated to prevent 73% of instances in persons with at least one underlying health condition. The vaccination has also been proven to induce equivalent immune responses in elderly persons when compared with young, healthy individuals.

The ChAdOx1 to-19 vaccination is administered as a two-dose course, which is delivered as an injection into the upper arm. The second dose is given 4-12 weeks following the first dose. In the UK, Pfizer-BioNTech or Moderna vaccines are then used for future booster doses, however, certain persons may be provided a booster dose of the Oxford-AstraZeneca vaccine if they cannot obtain the Pfizer-BioNTech or Moderna vaccinations.

Expected Side Effects

Because vaccinations operate by activating the immune system to generate a response, there might be side effects after taking the vaccine that seem similar to having a genuine virus. Things, like having a temperature, or feeling achy, or getting a headache (commonly characterized as "flu-like" symptoms), are typical after receiving numerous vaccinations and this is the same for the authorized COVID-19 vaccines. Having these symptoms suggests that your immune system is acting as it should be. Usually, these symptoms last a lot shorter than an actual illness would (most are gone within the first 1-2 days) (most are gone within the first 1-2 days).

The common adverse effects connected with the presently licensed vaccinations are below. These effects normally persist for 1-2 days after inoculation.

ChAdOx1 nCoV-19 (AstraZeneca-Oxford)	BNT162B2 (Pfizer-BioNTech) Range in brackets show difference between first and second doses
Arm pain (67%)	Arm pain (92-83%)
Chills (51%)	Chills (33-58%)
Fever (18%)	Fever (17%)
Joint pains (31%)	Joint pains (17%)
Muscle aches (60%)	Muscle aches (25-58%)
Fatigue (70%)	Fatigue (42-75%)
Headache (68%)	Headache (50-67%)

Vaccine Ingredients

Common chemicals used in vaccinations, and present in both the Pfizer-BioNTech and Oxford-AstraZeneca vaccines are: Sucrose (sugar) and acidity regulators such as Histidine, sodium, and potassium salts. The active element of the Pfizer-BioNTech vaccine is BNT162b2, which includes the genetic code for the coronavirus spike protein, within a lipid (fat) capsule. The vaccination also includes other inactive components such as cholesterol.

The active element of the Oxford-AstraZeneca ChAdOx1 nCoV-19 vaccine is developed from a modified adenovirus which causes the common cold in chimpanzees. This virus has been engineered such that it cannot create an infection. It is employed to transmit the genetic instructions for the coronavirus spike protein. The vaccine also includes inert substances such as polysorbate 80, an emulsifier, and a very little quantity of

alcohol (0.002 g per dosage) (0.002 g per dose). The vaccination also includes quantities of magnesium (3 to 20 parts per million) (3 to 20 parts per million).
All vaccine components are included in extremely minute concentrations and there is no evidence that they may cause damage in these amounts.

The Pfizer-BioNTech and Oxford-AstraZeneca vaccines do not contain:

- Human or animal products.
- Common allergens such as latex, milk, lactose, gluten, egg, maize/corn, or peanuts.

Both the Pfizer-BioNTech and Oxford-AstraZeneca vaccines employ genetic technology to produce an immune response.

The manufacturing procedure for the Oxford-AstraZeneca vaccine includes the manufacture of a virus, the adenovirus, which transports the genetic information to the cells within the body. To create this virus in the laboratory, a "host" cell line is required. For certain vaccinations, chicken cells are employed for this procedure, whereas other human cell lines are used to create the virus. The Oxford-AstraZeneca vaccine employs a cell line called HEK-293 cells. HEK-293 is the designation given to a particular line of cells employed in many scientific applications. The initial cells were extracted from the kidney of a lawfully terminated baby in 1973. HEK-293 cells utilized currently are clones of the original cells but are not actually the cells of the aborted fetus.

Other therapeutic products which utilize HEK-293 cells as a producer cell line include Ad5-based vaccines, such as Cansino's COVID-19 vaccine,

Adeno-associated viruses (AAV), and lentiviruses as gene therapy vectors for different disorders. Many of these items are in clinical testing.

Chapter 6: The Effectiveness Of COVID-19 Vaccines Against Omicron BA.5 and The Bivalent Booster.

COVID-19 vaccinations continue to give strong protection against the BA.5 variant. Vaccines are very useful in avoiding serious illnesses that may induce hospitalization.

Since the bivalent boosters, known as the "updated boosters" are so new, their efficiency is not entirely determined yet. However, prior research demonstrates that bivalent vaccinations result in a better immune response to contemporary variations such as BA.5 and help protect against the original strain. The bivalent vaccinations also expand the immune response, which is likely to boost protection against future variations and prolong the level of protection.

The bivalent booster vaccination is authorized by the U.S. Food and Drug Administration (FDA) and the Centers for Disease Control and Prevention (CDC) (CDC). It's an mRNA vaccine that targets the original SARS-CoV-2 virus and another component identified in both the omicron BA.4 and BA.5 strains. Moderna and Pfizer both offer a single-dose bivalent vaccination. Moderna's bivalent booster is offered to ages 18 and older, whereas Pfizer's bivalent booster is for ages 12 and older.

COVID-19 immunizations remain the greatest public health tool to help protect individuals against COVID-19. Vaccination may also assist to limit the risk of new variations arising. This covers main series, booster doses, and supplementary doses for individuals who require them.

Data is still being gathered on the vaccination efficacy against Omicron sub-variants including BA.4 and BA.5. However, vaccination efficacy against severe sickness and mortality remains strong for other Omicron sub-variants and presumably also for BA.4 and BA.5.

Research published in the CDC's Morbidity and Mortality Weekly Report also shows that during both Delta- and Omicron-predominant periods, receipt of the third dose of an mRNA vaccine was highly effective at preventing COVID-19-associated emergency department and urgent care encounters as well as preventing hospitalization.

Meanwhile, a JAMA research showed that receipt of three doses of an mRNA COVID-19 vaccine—compared with being unvaccinated or having had two doses—was related to protection against both the Omicron and Delta versions.

Chapter 7: What You Need To Know Before, During And After Receiving A COVID-19 Vaccine: Tips For Navigating The Vaccination Process.

Millions of individuals throughout the globe have now been securely vaccinated against COVID-19, making it simpler for the world to fall back into place. They are:

Before You Go:

1. Do your research:

There's a lot of misinformation regarding vaccinations online, so it's crucial to always receive information from trustworthy sources like UNICEF and WHO. If you have any doubts about whether you should obtain a COVID-19 vaccination, talk to your doctor. At present, patients with the following health problems should not get a COVID-19 vaccination to prevent any potential side effects:

If you have a history of severe adverse responses to any components of a COVID-19 vaccination.

If you are presently ill or having symptoms of COVID-19 ((although you can get vaccinated once you have recovered and your doctor has approved)

2. Talk to your doctor:

If you've ever encountered a serious adverse response from any vaccination or you have any concerns about the drugs you are presently taking, speak to your healthcare practitioner before your visit.

3. Take care of yourself:

Get a good night's rest and drink properly before your immunization so you can feel your best on the day.

During The Appointment

1. Stay safe:

Make careful to observe safety rules at the immunization facility such as physical distance while waiting and wearing a mask.

2. Communicate:

Let the health care expert know if you have any medical conditions that may be considered precautions, such as pregnancy or a damaged immune system.

3. Keep your records:

You should get a vaccination card that informs you of the COVID-19 vaccine you had, when you received it, and where you received it. Make careful to hang on to this card in case you need it in the future.

After You've Been Vaccinated

1. Stay for monitoring:

The health care practitioner should follow you for around 15 minutes after the vaccination is delivered to make sure you don't have any acute responses. However, it is quite unusual for serious health reactions.

2. Be prepared for some side effects:

Vaccines are supposed to provide you protection without the hazards of having the illness. While it's typical to acquire immunity without side effects, it's also common to suffer some mild-to-moderate side effects that go away within a few days on their own.
Some of the mild-to-moderate adverse effects you may encounter following immunization include:

- Arm discomfort at the injection site
- Mild fever
- Fatigue

- Headaches
- Muscle or joint aches
- Chills
- Diarrhea

If any symptoms remain for more than a few days or if you encounter a more serious response, then contact your healthcare practitioner immediately.

3. Be patient:

Building immunity takes time and some vaccinations need two doses. Make careful to examine how long it will take until you are deemed completely immunized.

Chapter 8: Do-It-Yourself (DIY): How To Be Safe

Protecting ourselves and our family members against the extremely infectious Omicron strain of COVID-19 is everyone's top concern right now. Workplaces are obligated by law to have several preventive measures in place and the most essential thing you can do is decrease your chance of exposure to the virus. The basic techniques to safeguard yourself and your loved ones are:

Getting your immunizations is incredibly essential. A double vaccination plus a booster strengthen your immunity against Omicron and considerably minimize your chance of being extremely sick or hospitalized."

Limit indoor gatherings. It's sort of a challenging thing to accomplish but it's vital. If you do have guests in your house, remain up to speed on provincial rules and maintain physical separation, respiratory etiquette, hand hygiene practices, and fast testing.

Another good technique is to increase ventilation in your house. Omicron is distributed mostly by respiratory droplets in the air. If you can dilute particles, you can help lessen the danger. Here is how:

- If you have ceiling fans, turn them on and make sure they are drawing air upwards.

Change furnace filter) (or ask your landlord or building manager if filters have been changed).

- Run your HVAC under the "on" function instead of "auto".
- Consider buying or renting a HEPA filter
- Turn on and leave on exhaust fans in bathrooms and cooktops when visitors arrive.

- Open windows or screen doors to enable fresh air flow.

Wearing medical masks is suggested. N95 or KN95 respirators are favored because they provide numerous levels of filtration. For an extra layer of protection, a medical mask may be augmented with a three-layered fabric mask. Make sure that your hands are clean as you put on and remove your mask. Fit is highly crucial. If you don't have a proper fit to your face (around your nose, mouth, and chin), the protection efficacy of the mask is substantially diminished. For medical masks, employ the "knot and tuck" method to optimize the fit. Tie knots in the ear loops, then tuck in the sides of the mask so that it sits flat on your face. Conduct a visual check in the mirror for visible facial seal gaps with the mask. If using a respirator, do a seal check with each usage following the manufacturer's directions.

You may also reuse disposable surgical masks, cotton masks, and N95 or KN95 respirators. Cloth masks may be cleaned; surgical masks, while normally disposable, may be placed into a brown paper bag for a few days to air out. Respirators may also be re-used for many days - per the manufacturer's recommendations. Replace the mask if it gets filthy, broken, soiled, or moist, or if you have been in touch with an infected individual.

If you have facial hair. It is virtually tough to acquire a decent fit with a mask or respirator. Consider shaving it off.

It's usual practice for employees who may be in close contact with another person to wear eye protection (goggles, face shields, safety glasses) as well as masks. This technique should also be considered by the general population. Eye protection may minimize pathogenic particles from being absorbed via the mucous membranes.

Getting a fast test now and again. A quick test gives an extra layer of protection, but a negative test doesn't guarantee that you are free and clear. You still have to employ additional precautionary measures. A positive test will warn you and your family to follow the isolation guidelines provided by the authorities. If you are confused about what to do, call your local public health unit.

Furthermore

- Physical distance of at least one meter from another,
- Regular handwashing, and
- Working from home while you're ill,

Can also aid in stopping the transmission of the virus.

So we're very much still in this epidemic. We need to continue to be watchful, need to be careful, and live our lives appropriately and securely.

www.ingramcontent.com/pod-product-compliance
Lightning Source LLC
LaVergne TN
LVHW020525160826
845677LV00015B/3908
* 9 7 9 8 3 5 9 4 6 3 7 9 9 *